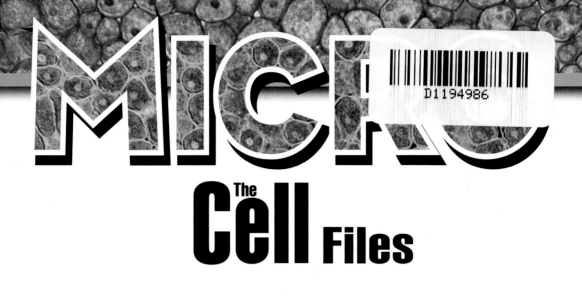

MICRO

The Cell Files

Discovery Channel School
Science Collections

DISCOVERY CHANNEL SCHOOL

© 2000 by Discovery Communications, Inc. All rights reserved under International and Pan-American Copyright Conventions.
No part of this book may be reproduced in any form or by any electronic or mechanical means, including
information storage devices or systems, without prior written permission from the publisher.
For information regarding permission, write to Discovery Channel School, 7700 Wisconsin Avenue, Bethesda, MD 20814.
Printed in the USA ISBN: 1-58738-002-1

1 2 3 4 5 6 7 8 9 10 PO 06 05 04 03 02 00

Discovery Communications, Inc., produces high-quality television programming,
interactive media, books, films, and consumer products. Discovery Networks, a division of Discovery
Communications, Inc., operates and manages Discovery Channel, TLC, Animal Planet, Discovery Health Channel, and Travel Channel.

Writers: Jackie Ball, Lelia Mander, Monique Peterson, Gary Raham, Darcy Sharon, Suzy Sensbach, Alicia Slimmer, Rachel Waugh, Sharon Yates. **Editor:** Lelia Mander. **Photographs:** Cover, nerve cells, ©SuperStock, Inc.; p. 2, cells, ©SuperStock, Inc.; girl and grass, ©Corbis; p. 3, Aborigine, ©SuperStock, Inc.; p. 4, muscle cells, Visuals Unlimited/©John D. Cunningham; p. 5, girl and grass, ©CORBIS, plant cells, ©Discovery Communications, Inc., rod & cone cells, ©Ron Boardman/Stone; p. 6, amoeba, ©Dwight Kuhn '86; p. 10, cells from sycamore stem, ©John D. Cunningham; p. 11, leaf cells, ©Ron Boardman/Stone; nerve cells, Visuals Unlimited/©Triarch; algae, Visuals Unlimited/© R. Calentine; p. 12, sperm on egg cell, Visuals Unlimited/©David M. Phillips, embryo cells on a pin, ©Dr. Yorgos Nikas /Photo Researchers, Inc.; p. 13, center and right, ©PhotoDisc; p. 15, Keone Penn & Dr. Yeager, © Ann States/SABA; sickle cell, ©Stanley Flegler/Visuals Unlimited; p.16, cork, ©Lester V. Bergman/COR-BIS; MICROGRAPHIA, ©Bettmann/CORBIS, Anton van Leeuwenhoek, ©Bettmann/CORBIS; p. 17, Hooke's microscope, ©Brown Brothers, Ltd.; p. 18, Louis Pasteur, ©Bettmann/CORBIS; p. 19, yeast, ©Lester V. Bergman/CORBIS; p. 20, Euglena, Visuals Unlimited/©David M. Phillips. Chlamydomonas, Visuals Unlimited/©Cabisco, para-mecium, Visuals Unlimited/©M. ABBEY, stylonchia, Visuals Unlimited/©A.M.Siegelman; p. 21, E. coli, Visuals Unlimited/©Fred Hossler; lung cilia, ©Science Pictures Limited/CORBIS; amoeba, ©Dwight Kuhn; p. 22, Gilda Radner, ©Michael O'Neill /Outline; p. 23, SNL still, ©Lynn Goldsmith/CORBIS; p. 24, man with binoculars, ©Miguel S. Salmeron/FPG International, p. 26, sloth, ©Wolfgang Bayer/DCI; p. 27, Aborigine, ©SuperStock, Inc.; p. 31, yeast, ©Lester V. Bergman/CORBIS; algae, Visuals Unlimited/© R. Calentine; Anton van Leeuwenhoek, ©Bettmann/CORBIS; p. 32, Rousseau painting, ©Private Collection/A.K.G., Berlin/SuperStock; all other photographs, ©COREL. **Illustrations:** p. 14, diagram of sickle cells in capillary, Christopher Burke. **Acknowledgments:** pp. 22-23, excerpts from IT'S ALWAYS SOMETHING by Gilda Radner. © 1989 Simon and Schuster. Reprinted with permission.

MICRO

Back to Basics

Cells are amazing. On the one hand, they are life at its most basic. Each cell has the ability to use and store chemical energy, and reproduce. Cells are also the building blocks of life, the fundamental units of which all life is built. On the other hand, cells are extraordinarily diverse. Some carry out one primary function, while others are very complex and carry out several elaborate functions. All are designed for a particular role, whether they are one-celled organisms or part of a group of similar cells. Let MICRO show you more: Come along with Discovery Channel on a fascinating tour of the wild and wonderful world of cells.

The Cell Files

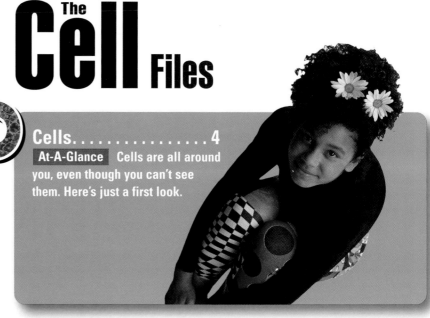

How does this musician get in the groove?
See page 26.

Final Project

Cells

Look around. Every living thing you see—every plant, every flower, every person, every bird or mammal—has one thing in common. Each is made of cells. Most of the organisms you can see are made of lots of cells. By lots, we mean LOTS. For example, you have about 50 to 75 million *million* cells you can call your own.

Even though you can't see most cells without a microscope, they are the basic building-blocks of all life. Believe it or not, most organisms on Earth are made of just one cell. The rest are like you: multicellular, or made of lots of cells.

All of the cells in complex living things work together to carry out particular activities. But that doesn't mean all cells are the same. Far from it! Plant cells and animal cells have basic differences, which you'll read more about at right and in the rest of this book. And even within the same kind of plant and animal, there is an amazing assortment of cell shapes and sizes. A cell is formed to do a particular job. Let's take a closer look.

BRIGHT AND BEAUTIFUL—Cells keep their organism alive. In addition, in people and in plants, some cells help the organism reproduce, so that life can continue. Different kinds of cells in a flower form the reproductive parts, which are needed to produce seeds for future generations of flowers.

BUILT TO MOVE—Plants generally stay in one place. They don't have to move around the way you do, so they don't need muscle cells. These long, stringy fibers can contract and relax. When a muscle cell in your body contracts, it's using energy to move some part of you. You know your muscle cells in your arms are working when you pick up something heavy. But they're also busy extending and contracting when you smile, frown, breathe, or laugh. Your heart muscle cells never get a break, keeping your heart beating even when you are asleep.

muscle cells

LAYERS OF CELLS—Plants have a layer of outer cells to protect what's inside, and so do you—your skin cells. Touch your hand and you're touching many thousands of skin cells. All of these outer cells are dead. But just underneath this surface are neat rows of living skin cells. Nestled among them are other types of cells: nerve cells, which allow you to feel things; fat cells to keep you warm; and capillary cells that make sure all cells are supplied with oxygen and nutrients from your bloodstream.

RODS AND CONES—Your eyes have nerve cells that convert light into electrical impulses and then send them to your brain. Cone-shaped cells pick up bright lights and colors, and rod-shaped cells respond to dimmer signals.

GREEN BLOCKS—Green plant cells turn sunlight energy into food. And plant cells have something that animal cells do not have: thick cell walls that give plants strength and structure.

Gotta Split

Interview with an amoeba

Q: You're an amoeba, composed of only one cell. A kind of shapeless cell, if I may.

A: No, you may not. I happen to like my shape very much. And besides, I can change it any time I want, and fast (well, more or less).

Q: Well, yes, but no matter what you do, an amoeba's shape is still kind of blob-like. Maybe you should consider getting more exercise. Firm up that jelly-like stuff a bit.

A: MORE exercise? I spend my life exercising. That's all I do. Only I prefer to call it work. Hard work.

Q: Hard work? How can something that looks so soft and squishy work hard?

A: Because all cells have to have food energy—yours, mine, an artichoke's. And in my case, eating means I have to move. Hunt. Sneak up on some unsuspecting piece of pond scum or clueless bacterium. Unlike certain multicelled organisms that get to stay in one place and make their food, I have to go out and get my chow myself.

Q: Which multicelled organisms are you talking about?

A: Plants. You know, the green things that stay in the same spot their whole lives? Plants have it made. Every time the Sun shines, the dinner bell rings.

Q: How so?

A: Plants don't have to go anywhere to get their food. That's what being planted means: staying in one place. Even plants that "creep," like vines, are rooted in one place that doesn't change. The point is, everything plants need to make food is either built into their cells or always within reach. They take in carbon dioxide from the air. They slurp water out of the ground. Then sunlight helps to change it all into food right in their own leaf cells. It's not fair.

Q: But it sounds like a good trick. Can people learn to do that? Manufacture food right in their cells?

A: Nope, but that's why humans created supermarkets, take-out pizzas, and lots of other food suppliers. Each person has about fifty to a hundred million million cells. That's a lot to feed. You need a lot of food. You also need something else.

Q: What?

A: A mouth. The first stop on the road to each and every individual cell in your body. Of course, that's only the beginning. Food needs to get mashed and strained and pushed through your digestive system before it can reach your cells.

Q: Once it gets there, how does the food actually get inside our cells ?

A: One word: membrane. That's the layer on the outside of a cell. All cells have them. Me, too! Membranes are designed so that nutrients can move in and wastes can go out. That's how your cells get fed and that's how my cell—which you can also call my "self"—gets fed. I love my membrane. I use it in a really special way.

Q: Which is?

A: Since it's nice and flexible, I can fold it all around the bacterium (or alga or whatever is my menu of the moment) and then pull my prey inside me whole. Then I digest the parts I want and push the waste out later.

Q: How often do you do this folding and eating thing?

A: Constantly. Never stop. Love to eat. Live to eat. I eat and eat and eat until I've doubled my size. And then something wonderful happens.

Q: What is it?

A: I divide into two brand-new cells. That's my purpose in life—making more of me. Making two independent-minded, free-thinking, wonderful one-celled creatures where there was just one. It all happens in only about a day—just twenty-four hours.

Q: Pretty impressive. Do all living things reproduce like you?

A: Nope. All cells have to divide for their organisms to grow. But a multicelled organism needs a partner to reproduce. We solo cells only need ourselves. Our type of reproduction is asexual, and it's called binary fission. Speaking of which, we'd better wrap this up. I need to split in a few minutes.

Q: OK. Last question. When all's said and done, would you ever want to trade the stressful life of a short-lived, constantly hunting one-celled organism for the slower-paced life of a multicelled creature? Say, a human being?

A: No way. I like things the way they are: just me, myself, and I. No complicated organs and systems and such. Give me the simple life. And I have lots of company. There are way more one-celled organisms in the world than multicelled ones. My ancestors were here at the beginning of life on Earth, or at least a long time ago. Who knows? Someday we might just be the only ones left.

Activity

IN AND OUT An amoeba's form is bounded by its membrane. The membrane shape changes as the materials inside the amoeba shift around and push against the membrane. The main ingredient inside a cell is water. A main ingredient of a cell's membrane is a layer of lipids, or fats. The layer of fat provides a clear boundary from the water. To see how this works, fill a glass about halfway with water, and add a drop of food coloring to the water. Then add about half a cup of cooking oil, such as olive oil or vegetable oil. Stir the mixture vigorously and then let the contents settle. Write down your observations.

ALMANAC

You and Your Cells

Cells are the building blocks of all multicelled organisms—including you! They make up your bones, skin, organs, and most other parts of your body. Right now, you have over 50 million million (that's 5×10^{13}) cells in your body, and each one is busy doing its job.

LOOKING INSIDE A CELL

Even though cells are the tiniest parts of our body, we know a lot about how they work. Let's take a peek inside the average cell and see what it's made of.

■ The outer layer of every cell is called the **MEMBRANE**. It acts as a gate, letting supplies in and waste products out.

■ Every cell contains **CYTOPLASM** within its membrane. This is a jelly-like substance that helps support all the parts inside the cell.

■ Cells make energy out of food in their **MITOCHONDRIA**. This energy fuels all cell activities. (Plant cells also have **CHLOROPLASTS**, bodies that use sunlight to make food, in a process called **PHOTOSYNTHESIS**.)

■ A cell's "brain" is the **NUCLEUS**. This unit contains all the cell's genetic information and also tells the cell what to do. Bacteria do not have nuclei.

THE STUFF OF CELLS

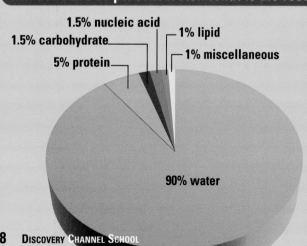

All cells are 90 percent water. What is the rest?

1.5% nucleic acid
1.5% carbohydrate
1% lipid
5% protein
1% miscellaneous
90% water

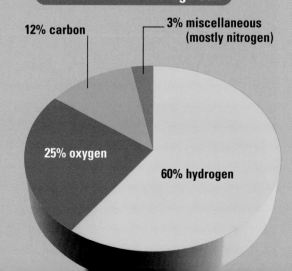

Elements in an average cell

12% carbon
3% miscellaneous (mostly nitrogen)
25% oxygen
60% hydrogen

SHAPES AND SIZES

There is a direct link between a cell's job in your body and how it is shaped.

BONE cells aren't all hard and brittle, as you might think. Like your other cells, they are soft and surrounded by membranes. But these cells acquire calcium and are grouped to form bony tissue.

FAT cells look like a bunch of bubbles. They store energy and help insulate the body from cold.

SKIN cells are shaped a little like building blocks, all fitting neatly together. They form your skin, which holds all of you together!

LIFE AND DEATH

In the time it will take you to read this chart, about 300 million cells will die inside you. The good news is, while those are dying, about the same number of new ones are being created. But not all cells die at the same rate. You are made up of 200 different types of cells. Some will last a long time, others will have very short lives.

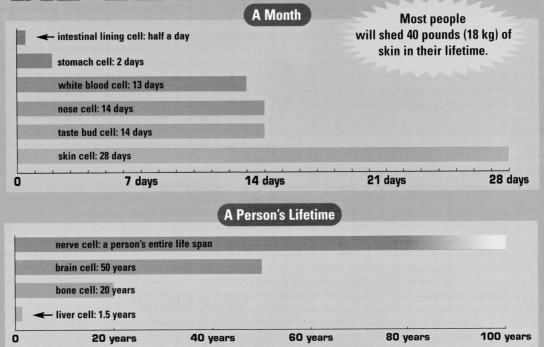

A Month

- ← intestinal lining cell: half a day
- stomach cell: 2 days
- white blood cell: 13 days
- nose cell: 14 days
- taste bud cell: 14 days
- skin cell: 28 days

0   7 days   14 days   21 days   28 days

Most people will shed 40 pounds (18 kg) of skin in their lifetime.

A Person's Lifetime

- nerve cell: a person's entire life span
- brain cell: 50 years
- bone cell: 20 years
- ← liver cell: 1.5 years

0   20 years   40 years   60 years   80 years   100 years

"EYE" CAN SEE YOU!

The retina of the eye contains about 137 million light-sensitive cells in an area about 1 square inch. There are 130 million rod-shaped cells for black-and-white vision, and 7 million cone cells for color vision. Do the math: How many rod cells do you have? How about cone cells? How many in your immediate family?

Activity

KEEPING TRACK Refer to the chart on different life spans of cells and think of your own body. Make a chart of the past month and plot which of your cells—intestinal lining cell, stomach cell, white blood cell, nose cell, taste bud cell, and skin cell—lived and which died during that period. Plot one of each type on your chart. Add other sections to the chart, and fill those in with your daily activities. Note how much time you spent sleeping, eating, going to school, and so on. What was going on with your different cells during that time? Which cells are still around, and which have been completely replaced?

Formed for

Cells are the smallest units of every kind of life—they are the building blocks of all life, from food mold to elephants to everything in between. So you'd be right if you think there are lots of different kinds of cells.

Think of cells in a multi-celled organism as members of a baseball team. Each one has a specific task on the field, so that the players can perform their best as a team. The pitcher has a certain set of skills for throwing the ball, which are different from the skills a good catcher needs. Cells are set up the same way. They are built differently and have characteristics based on their purpose. Some have little hairs because they must catch something passing by. Some have extra thick walls to support weight. There are as many differences among cells as there are tasks. Here you'll meet some cells as different as they can be.

Cells for Support   Xylem Cell

Without our bony skeleton to hold us up, we'd sink to the ground in a baggy heap. Plants don't have bone cells, but they have cell walls to support each cell. Some have a second cell wall for extra support. For example, xylem (ZY-lem) cells in woody trunks are tube-shaped and have extra thick cell walls to help a tree grow and support its weight. This is a cross section of cells from the stem of a very young sycamore tree. Wood cells at their maturity lose their internal contents, and then thick cell walls form.

Function

Cells for Protection
Leaf Epidermis

Our skin protects the important stuff inside our bodies from the outside world. Therefore, skin or any exposed cell has to be tough. That goes for both animals and plants. People's skin cells fit together neatly, like bricks in a wall (see page 9). The outer cells on plant leaves, called epidermis cells, fit together like pieces of a jigsaw puzzle. A layer of wax coats each plant cell exterior. The tight fit and waxy coating keep out dangerous germs and foreign objects and help regulate water loss.

Single-Minded:
One-Celled Organism

Some cells live independently. These are called single-celled organisms. The cell shown below is a green alga from a pond. Algae cells have a green pigment called chlorophyll, which allows the cell to make its own food from sunlight using carbon dioxide and water.

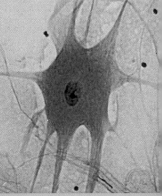

You Have a Nerve:
Cells for Communication

Nerve cells (example at right) carry electrical messages to and from different parts of the body. Their long extensions, called axons, add to the cell's length. The axons can help carry the messages long distances. One nerve cell controlling a human's arm and leg muscles can be more than 3 feet (1 m) long!

Oh Relax:
Cells for Movement

Many animals have long, thin muscle cells that allow them to move. These cells overlap one another and slide back and forth to expand or contract a muscle. Some muscles move voluntarily, such as those used when walking or lifting. Other muscles move involuntarily, such as those you use to digest food and breathe.

All Systems . . . GO!

What happens as you grow? You get taller, your feet get bigger—but does this mean your cells get bigger, too? The answer is no. Your cells stay a certain size, but as you grow they divide and multiply, creating more cells. The bigger you are, the more cells you have. It's as simple as that. But how did it all start?

Although we have many different kinds of cells—bone cells, muscle cells, neurons, lymphocytes—they all have one thing in common. Each one is a descendant of our very *first* cell: a zygote, an egg fertilized by a sperm. What started out as one cell becomes many cells. At first, these are all exact copies of the zygote. But at a certain point this changes—identical cells divide into different kinds of cells. Each one is programmed in a special way to perform its task and create new cells of its own kind. Let's take a look.

Day 1	Days 2-5	Days 6-8	Week 2
The zygote is formed inside the mother's body when a male sperm cell, from the father, makes contact with the mother's egg cell. Chemicals from the sperm dissolve a spot in the egg's protective membrane so it can enter. The zygote soon divides into two identical cells.	On the second day, the two cells split again, becoming four identical cells. On the third and fourth days, the cells divide again into 8, then again into 16 cells. The cells continue to double in number.	The cluster of cells, called a blastocyst (meaning "ball of cells"), is made of about 100 cells by the end of the first week. These cells are being fed by nutrients in the original egg cell. The blastocyst then anchors itself in the uterus's lining, which is rich in nutrients to feed the new cells. At this point, they separate into two groups. The inner group will become the embryo. The outer cell group will become the placenta, through which the embryo will get oxygen and nourishment.	Until now, the embryo's cells are identical. But at a certain point between the second and third week, the cells divide and form different *kinds* of cells. These cells look different and are programmed to do different things. Once a cell becomes specialized, it stays that way. When it divides, it can only create new cells of that same type. It can't change back to a general cell or become another type of cell.

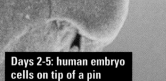

Days 2-5: human embryo cells on tip of a pin

Cells build all the organs and systems we need before birth, so we can grow from babies to kids, and from kids to adults.

Week 3

Three distinct layers form in the embryo. The top layer is the beginnings of the nervous system. Cells in this area will form the brain, spinal cord, and backbone. The embryo's circulation system starts taking shape in the middle layer. These cells begin building the heart and blood vessels, and dividing into red and white blood cells. The cells in the third layer, closest to the placenta, build a simple tube that will one day grow into the respiratory and digestive systems.

Days 21-22

The fetus is only $\frac{1}{17}$ of an inch long, but it's growing rapidly. As more and more cells form, they need a network for receiving nutrients and carrying away waste. The heart beats for the very first time. Circulation comes first: No other tissues, organs, or systems can grow until all the fetus's cells can be properly nourished through the blood.

Weeks 4-40

The heart is now pumping blood. From this point up until birth, all the organs and systems will continue to develop and grow. The fetus receives oxygen and nutrients from the mother's placenta through the umbilical cord. At birth, the baby has all the tissues, organs, and systems needed to grow into a kid, a teenager, and then an adult. And to think it all began with that very first cell!

Activity

SEE HOW THEY GROW In the early stages of life, growth is key: If the very first cells don't divide properly, the blastocyst won't grow into an embryo. Using 100 pennies, compare what happens when something goes wrong. First, start with one penny, then add another to represent the first cell division. Add two more, for the second division, so that your total is four. Keep adding enough pennies, as each "cell" divides into two new ones. How many cell division stages does it take to use up the pile of 100 pennies?

Now imagine that after the first division, one cell kept dividing and the other didn't. Using the same number of division stages as in the first exercise, see how many pennies you wind up with. How many are left over? Put your results in two different tables and compare them.

Cells to the Rescue

In a daring new procedure, doctors operated on a 12-year-old boy named Keone Penn. Since birth Keone had suffered from a disease called sickle cell anemia. This particular operation could cure him for life, or it could fail. No one knew for sure what would happen. At a time when most teenagers are thinking about sports, dating, movies, or Nintendo, Keone Penn was thinking only about survival. "I think about it a lot . . . I can't let it travel through my mind every day without saying, 'Lord, I need your help,'" said Keone.

When Blood Cells Go Wrong

To understand what sickle cell anemia is, you first have to know about red blood cells—which aren't technically cells at all. Instead of having a nucleus, they're filled with hemoglobin, a molecule that carries oxygen throughout the body. Healthy red blood cells are round. Their shape lets them slip easily through tiny blood vessels.

But a "sickle" red blood cell is shaped like a crescent, or sickle, a tool used to cut down fields of wheat. Sickle cells can't easily pass through blood vessels. As a result, the cells often get stuck, blocking the capillary openings. Other red blood cells can't get through either, so they can't carry oxygen to the cells nearby.

People with sickle cell anemia can suffer all kinds of serious complications, including strokes (when blood can't get to the brain), lung congestion, and kidney damage. The disorder is often accompanied by extreme pain throughout the whole body.

One Boy's Long Road

More than 80,000 Americans suffer from sickle cell anemia. Many of these are African Americans; every year, one out of every 375 African-American children is born with the disease. Keone Penn is one of them. Since he was five years old, Keone has had to have monthly blood transfusions. He also suffered from constant pain, fevers, infections, and seizures. It seemed his time was running out: Keone's family and doctors worried that he would have a stroke that could kill him. The only hope for Keone was a bone marrow transplant.

In this procedure, a patient's blood-making factory, the bone marrow, is replaced with marrow cells from another person, or donor. If everything works, the patient's body will make a whole new system of red and white blood

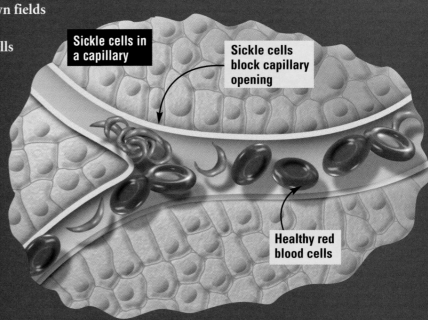

Sickle cells in a capillary

Sickle cells block capillary opening

Healthy red blood cells

cells. Usually the donor is a family member who has closely matching bone marrow cells. Keone didn't have relatives whose cells matched, so Dr. Andrew Yeager and the medical team at Emory University Department of Pediatrics decided to try something new. The doctors took cells from the umbilical cord of an unrelated newborn baby. Yeager used these particular cells because they can divide and re-divide indefinitely, and they have the potential to develop into any kind of cell.

Risks and Rewards

A brave Keone underwent 10 painful days of chemotherapy to kill his own bone marrow, which was producing the sickle cells. Then, Dr. Yeager performed the groundbreaking transplant at Eagleton Children's Hospital in Atlanta, Georgia. Other sickle cell patients had been cured with bone marrow transplants from related donors. But this type of cell transplant had never been performed. The doctors believed there was a 50-50 chance the procedure would fail. There was a high risk that Keone's body would reject the new blood, or infections could set in. Both prospects were dangerous and could even be fatal. But Keone was hopeful. "If the treatment works, I want to play football," he said.

To everyone's relief, Keone pulled through the operation, and his body accepted the new cells. Since then his health has improved dramatically. His family and doctors hesitate to use the word "cured," but so far, there has been no trace of the disorder. "The blood cells are now fully operational, making all healthy blood cells. We see no sign of sickle cells," says Dr. Yeager. Keone couldn't be happier. "I made it," he says. "My heart's beating and my brain is working. So much to be thankful for."

Keone Penn with Dr. Yeager

Sickle cell

Healthy red blood cells

Activity

TINY TUBES Blood delivers nutrients and oxygen to the body's cells through the smallest vessels in your blood stream, called capillaries. Some capillaries are so narrow that only one red blood cell can pass through at a time. To see how disruptive sickle cell anemia can be, log onto this Web site: http://www.fhcrc.org/about/CenterNews /pubrel/CNews1996/Aug15/Multicenter.htm and observe the animation. Write down your observations and conclude with a description of how this disorder harms tissues of the body.

Seeing the Invisible

O ne of the hardest things about studying cells is that they're usually too small to see with the naked eye. For the longest time scientists didn't even know there were such things as cells. About 400 years ago, the microscope appeared on the scene.

Cork cells enlarged more than 5,000 times

London, 1665

Hooked on Cells

Robert Hooke changed the scientific world forever by publishing *Micrographia*. This book included his drawings of objects he had observed under a microscope he had built. Hooke described looking at magnified pieces of cork. He saw rows and rows of little boxes, which reminded him of the cavities in a bee's honeycomb. He called them cells, meaning small chambers. Picture a cell in jail.

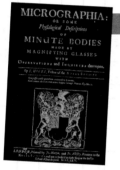

Leeuwenhoek

Take a Closer Look

The magnifying glass played a big part in the history of microscopes when fabric merchant Anton van Leeuwenhoek (pronounced "LAY-ven-hook") opened a fabric store in Holland. He used a magnifying glass to examine cloth, and before long Leeuwenhoek began building simple, or single-lens, microscopes that magnified objects 270 times.

Delft, The Netherlands, 1674

"Animalcules"

Leeuwenhoek had always wondered why the water in the lake near his home turned cloudy every summer. He examined the water under his microscope and saw tiny creatures swimming around—tinier than any creatures ever seen before. He called them animalcules and described their colors and movement:

" ...the motion of most of these tiny creatures in the water was so fast, and so random, upwards, downwards, and round in all directions, that it was truly wonderful to see."

Scientists later called these creatures microorganisms.

Power Vision

Optical microscopes, like the ones you use in school, magnify objects by bending light that passes through the lens. Electron microscopes work in a similar way, but they use a beam of electrons focused by a magnetic lens. The difference in results is staggering.

Instrument	Number of times object is enlarged
Magnifying glass	10–20 times
Average optical microscope	40, 100, and 400 times
Advanced optical microscope	2,000 times
Electron microscope	1 million times

Small, Smaller, Smallest

You can see some plant cells without even touching a microscope. The human eye can see objects as small as 100 micrometers. And that's the size of some plant cells. A micrometer is one millionth of a meter. In other words, 1 million micrometers make up a meter. The symbol for a micrometer is the Greek letter µ ("mu") and m.

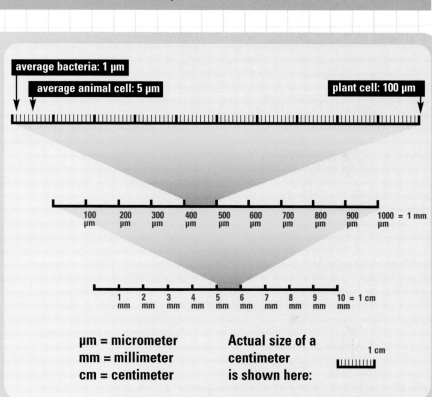

µm = micrometer
mm = millimeter
cm = centimeter

Actual size of a centimeter is shown here:

1 cm

Drawing from Life

Robert Hooke was trained as a painter. He used his artistic skills to make detailed drawings of the cells he observed through his microscope (below). Known as a compound microscope, this instrument used two different lenses and magnified objects 40 times.

Hooke's compound microscope

Activity

SIMPLIFY! The term "one millionth of a meter" is easier to grasp when you express it as a power of 10: 10^{-6} m. It's also a way to give a small measurement without using the µm symbol, which might not exist on every computer. Write the following cell measurements as powers of 10, and in terms of meters:

1) Fertilized human egg cell: 70 micrometers
2) White blood cell: 5–8 micrometers
3) Length of a nerve cell in your arm: 500,000 micrometers
4) Length of *Bacillus fragilis*: .5–1.5 micrometers

Answers on page 32

It's ALIVE!!

Something was wrong with the French winemaking industry. Some batches had spoiled, and no one understood why. Emperor Napoleon III invited Dr. Louis Pasteur, a professor of chemistry, to look into the problem. He chose the right man: Pasteur had been investigating the process for making wine and beer for several years and had some surprising answers.

Basically Biological

People had been making wine and beer for centuries, using a process known as fermentation. This was how juices from fruits and vegetables such as grapes and barley became wine and beer. Fermentation is also what makes bread rise. Most scientists at the time assumed fermentation was a chemical process of some kind. But Pasteur proved that tiny living things caused fermentation. Certain one-celled organisms absorb sugars and other nutrients through their membranes. After breaking down these substances, the organisms give off various substances as waste products—one of which is alcohol.

Dr. Louis Pasteur

Pasteur had been investigating the process of fermentation since 1854. By examining samples of beer under a microscope, Pasteur observed small, round particles in the liquid. In the spoiled beer, these were oval-shaped, while in the unspoiled beer they were shaped more like spheres. After more investigations and tests, he concluded that these particles were actually microscopic organisms. Fermentation took place when these organisms were healthy and able to reproduce. But if conditions were unhealthy—if, for example, the temperature was too hot—the microorganisms would not function properly and fermentation would fail. As a result, the beer or wine would be spoiled.

Through his experiments, Pasteur revolutionized the scientific community. He had proved that fermentation was a chemical process and that it was a biological one as well.

"Life Without Air"

Like all living things on Earth, cells need chemical energy to perform their various functions. Cells use chemical energy to process nutrients, to expel wastes, to grow, and to reproduce. This energy comes from nutrients absorbed and broken down by the cell, a process that for many organisms also requires oxygen. Just as we need oxygen to survive, our individual cells need oxygen to carry on their functions. And many one-celled organisms need oxygen for the same reasons.

Fermentation is one way a cell can get energy without oxygen. Pasteur observed one-celled organisms functioning and reproducing in environments without oxygen and concluded: "Fermentation is the consequence of life without air." Before this discovery, no one believed that life could survive without oxygen. Pasteur opened up a whole new world of investigation by proving that many forms could, through the process of fermentation.

A cell gets energy through fermentation by converting

nutrients such as sugars into usable chemical energy, but without using oxygen. The process gives off various waste products, which the cell disposes of through its membrane. Waste products vary, depending on the organism. Some yeasts give off carbon dioxide, which causes bread to rise. Others produce ethyl alcohol, which is the alcohol in beer and wine. And our muscle cells use fermentation, too, when they need extra amounts of energy. Ever notice how your muscles sting and ache after you've been exercising very hard? This is caused by lactic acid, a by-product of the fermentation taking place in your muscle cells.

The Germ Theory

Pasteur's work with fermentation led him to form one of the most important theories in the history of microbiology. His discovery that life could exist without air led him to examine microorganisms in a completely new way. If one-celled organisms were the cause of fermentation, couldn't they also be the cause of disease? Until then, people didn't understand that infections were caused by microscopic life-forms. Pasteur identified several disease-causing bacteria and demonstrated how these could be killed by heat. The process he recommended is still used to kill harmful bacteria in

milk. Today we call it pasteurization. Pasteur's work also led to the sterilization of medical instruments, which greatly reduced infections in operations.

Pasteur devoted his work to understanding life on the smallest scale. Although he completely revolutionized the fields of industry and medicine, Pasteur always put his faith in the scientific process. "Imagination should give wings to our thoughts," he once said, "but we always need decisive experimental proof."

In the background picture is a magnification of a one-celled fungus commonly called baker's yeast. It's waste product, carbon dioxide, causes bread to rise.

Activity

YEAST IN ACTION You've learned that yeast organisms cause bread to rise, but you can't easily see it as it's happening. Try this experiment to observe the by-products of fermentation. Take five glass bottles, and label and fill them halfway with the following solutions:

- Bottle #1: **Water**
- Bottle #2: **Sugar-water** (1 part sugar to 4 parts water)
- Bottle #3: **Starch-water** (1 part mashed potato flakes to 4 parts water)
- Bottle #4: **Sugar-protein water** (1 part beef broth and 1 part sugar-water)
- Bottle #5: **Protein-water** (beef broth diluted to half-strength with water)

Then add a pinch of baker's yeast to each bottle and cover each bottle opening with a balloon. Record the appearance of each solution and then leave the bottles at room temperature.

On the next day, record the appearance of each solution and make a table of your observations. The table should indicate changes in appearance from the first to the second day, and any differences between the five solutions. Write a brief report based on your observations. What do the different solutions reveal about yeast and fermentation?

Going Solo

Billions of cells are swimming, crawling, or just drifting along through life. One-celled organisms tend to move around a lot. One type of seafaring bacterium uses air sacs to float up to the ocean's surface. Another bacterium produces a layer of slime to slither along—just like a snail. So why do one-celled organisms move? For the same reasons most other animals move: to escape danger, and to find food. Sometimes they need to travel from a hostile environment to a place where they will be able to survive. And if they don't move fast, they'll die. There is a great variety of one-celled organisms, and one way to study this diversity is to look at the many different ways they can get from one place to another.

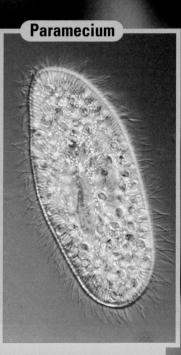

Paramecium

Wagging the Tail

From a microscopic point of view, the fluids some one-celled organisms travel in are as thick as molasses, and just about as easy to swim in. Many use the tadpole approach, wiggling their tail, or flagellum, as if repeatedly cracking a whip. This is how *Euglena* (below) moves forward. The single-celled chrysomonad wriggles to move backwards. An alga, *Chlamydomonas nivalis* (above), uses its two flagella-like arms to swim "breaststroke" style. If it has to make a getaway, its "arms" coordinate their movements like a tail, propelling the alga quickly away from predators.

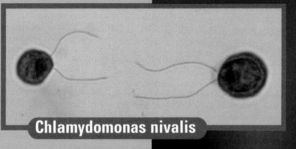

Chlamydomonas nivalis

Keeping Time

Some organisms, such as *Paramecium* (above), are covered with short "hairs" called cilia (which is Latin for "eyelashes"). These cilia beat like the oars of a small Viking ship, at a rate of about 20 beats per second. The cilia don't beat exactly in time with each other, creating a ripple effect that causes the cell to spiral forward like a slow, hairy football. In one type of cell, the complex protozoan *Stylonchia* (right), the cilia fuse together into stumpy little legs, called cirri. The cell can use cirri to "walk" over surfaces or even jump in emergencies.

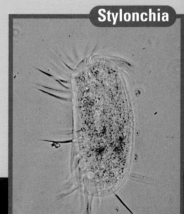

Stylonchia

Euglena

The Twist

Some bacteria, such as *E. coli*, have a bunch of flagella. They spin at the base like the blades of a propeller. The corkscrew tails are attached to a tiny "wheel" powered by a structure inside the cell. When rotating counterclockwise, the flagella spiral out behind the cell, smoothly pushing it along. The bacteria can switch to a clockwise rotation, which flings the flagella every which way.

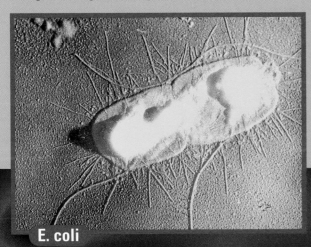

E. coli

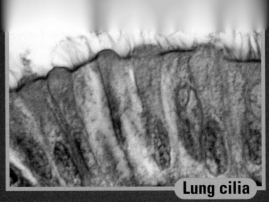

Lung cilia

Doing the Wave

Even when a cell equipped with cilia doesn't move, its cilia continue to undulate like a crowd doing the wave at a baseball game. This action moves liquid near the cell's surface and allows the cell to feed itself by washing particles into its mouth. Some human cells have cilia that move fluid around the brain. The cilia in your lung cells work to keep your lungs clean. Billions of cilia pass particles of dust and mucus from the lungs up to the throat, where they can be swallowed and later disposed. Cigarette smoking damages cilia, so smokers' lungs stay dirty.

Putting a Best Foot Forward

Some single-celled organisms use pseudopodia ("fake feet") to crawl. An amoeba will rearrange itself and protrude on one side. This blob will swell and extend forward. Then it grabs hold of the surface beneath it and pulls the rest of the cell after it. The fake foot then gets reabsorbed into the rest of the amoeba.

Amoeba

Activity

OBSERVE AND SKETCH See how flagella work! Hold a piece of string or rope in your hand and flick your wrist several times. Watch the rope movements closely, then sketch the wave patterns on a piece of paper. Can you break up an average wave movement in a series of sketches? Now, make a flip book showing how a flagellum moves an organism forward. Take a pad of paper and draw an oval-shaped bacterium with a flagellum at the edge of the page. In the same spot on the next page, draw the bacterium with the flagellum in a different position, based on your sketches of the wave patterns you made earlier with string. On the third page, draw the same bacterium but moved slightly to the right, with the flagellum at another stage of its motion. Continue drawing like this 15 to 20 pages. Flip the book pages rapidly and watch your bacterium go!

A FIGHT to the DEATH

Gilda Radner with Sparkle

blood system. Cancerous cells might spread to other parts of the body. If so, cancer causes illness and sometimes death.

One way to fight cancer is to attack the cancer cells directly. This is done by radiation, focusing high-energy beams onto the cancer cells. Another way is through chemotherapy, in which toxic chemicals are injected into the body. These chemicals target and kill cancer cells without doing too much damage to surrounding healthy tissue.

Gilda Radner, a famous comedian and actress, who appeared on "Saturday Night Live," discovered at the age of 40 that she had cancer. Over the next two and half years, she fought a tough battle. In the end, cancer won the fight and Radner died just before her 43rd birthday. But she didn't give in easily. Instead, she used every ounce of her humor and spirit to share her experiences with others involved in the same struggle. Here is her story.

Bad News

After experiencing several mysterious symptoms, Radner went into the hospital for tests. She and her husband, actor Gene Wilder, waited together to find out what was wrong.

We both looked up into this doctor's eyes as he said, very calmly, "We've discovered there is a malignancy." A flush went through my body, and out of my mouth came a sound like a guttural animal cry … Surgery would have to be done as soon as possible. When he left the room, I grabbed Gene's face in my hands and sobbed.

Normal cells grow and then divide, creating copies of themselves. This is how damaged tissue is replaced, for example, and how we grow from children to adults. Cancer occurs when one cell gets a message and divides rapidly. Instead of maturing so that it can do a specific job, a cancer cell is programmed only to divide. Cancer cells keep dividing, forming a mass of tissue called a tumor. The tumor keeps growing and growing, and before long it's interfering with the proper functioning of nearby organs. For example, cancer cells that start in the lungs will interfere with how lungs breathe in and out and pass oxygen to the

Something to Laugh About

Chemotherapy treatments began as soon as Radner recovered from the surgery. Even though all the visible signs of cancer had been removed in the operation, there was still a chance that microscopic cancer cells remained. These could quickly grow into dangerous tumors if they weren't stopped. Radner was grateful for the help of Joanna Bull, a gifted therapist who helped her through the worst parts of "chemo." Humor also was key.

Joanna taught me … that cancer cells are remarkably stupid. They are just the dumbest things you could imagine. They are like the guys in the foxhole who are supposed to be hiding from the enemy but who stand up and say, "I can't take it any more," and then get shot. When they see the chemo coming they run out and yell, "I'm here, I'm here." Your normal cells get hit by the chemicals and they are jolted, but they aren't stupid—they are smart and they say, "I'll just get myself back together again." That's why chemotherapy works, because of how stupid cancer cells are.

Finding Support

Another helpful source was The Wellness Community, a support group for cancer patients in Santa Monica, California. Through meetings with this group, Radner learned that there was a way to have a full and joyous life, even with cancer.

The Wellness Community … reminded me of the early days of "Saturday Night Live" when we … believed in making comedy and making each other laugh. We were just working together to entertain, like kids playing together.

The hardest part … was learning later that someone who had become close had died. The course of cancer isn't always what we hope. I was learning that death is a part of life. But if I hadn't gone to The Wellness Community, think of all the love I would have missed. While we have the gift of life, it seems to me the only tragedy is to allow part of us to die—whether it is our spirit, our creativity, or our glorious uniqueness.

Gilda Radner with Dan Ackroyd on the set of "Saturday Night Live"

Postscript

In October 1993 Gene Wilder opened Gilda's Club in New York City, where people suffering from cancer, and their loved ones, can gather for support, community, and information. Joanna Bull was the club's first executive director.

Activity

DIVIDING TO CONQUER To get an idea of how fast cancer cells can crowd out healthy tissue, try this exercise. Gather together a large number of two different types of objects, for example pennies and dimes, or two different sizes of paper clips. One type of object will represent healthy cells, and the other will represent cancer cells. Start with one object from each group, and using a second hand on a watch, demonstrate what would happen if in a certain type of tissue normal cells divide once every 10 seconds, while cancer cells divide twice as fast in the same amount of time. Cluster the objects representing the cancer cells in the middle of the objects representing healthy tissue. What has happened after one minute of cell division? What does this demonstration show you about how cancer cells can intrude upon healthy tissue cells? What might this mean for the organ that the healthy cells belong to?

Life in the Cellular Zone

To the naked eye, the world looks different depending on whether you view it from high in the sky or down on the ground. But if you could make yourself small enough, you'd see that it's all made of cells—different kinds of cells in different animals and plants, all designed to carry out special activities. Take this virtual voyage and see for yourself. . .

You start out in a field, looking up at a hawk with your binoculars. Suddenly you've become a speck, tiny enough to . . . be swooped up as the hawk dives. You look for a safe place on the tough surface of one of its talons, which is covered with scales. You work your way between them.

Along the way, you pass skin cells: hundreds of them, all looking alike. Each one has a membrane and a darker area you recognize as the nucleus. Some cells are dividing. You watch as the nucleus of one cell appears to dissolve, and the chromosomes form. These pull apart to opposite sides of the cell, and then a membrane forms in the middle of the cell, splitting it in half. Then a nucleus forms in each cell. What was one cell is now two cells.

You go deeper until you get to a different zone entirely. You're among muscle cells now. Unlike the skin cells, which fit snugly like puzzle pieces, these cells are long and bundled together. Each one contains bunches of fibers. Without warning, the fibers contract—you jump back as the muscle cells suddenly shorten. What's going on? You hurry back through the skin cells to the talon's outer surface.

Wow—what a change! The hawk is flying much closer to the ground. The prairie grasses are just several feet below you,

and then you're flying over a patch of brown turf. It's a prairie dog town—you can tell by the series of mounds that mark the entrances to their network of tunnels. The hawk's talons are flexed and he's getting closer and closer to his prey, a lone prairie dog running frantically for safety.

From Predator to Prey

The hawk strikes with a clenched talon. A clump of fur gets caught for a second, but that's all. The prairie dog has escaped! The hawk changes direction quickly. Meanwhile, you've lost your grip on the talon. You're falling, falling . . . and then you land on the back of the "lucky dog."

You grab a strand of hair and slide down. At its base the strand rises above you as big as a redwood tree. You burrow into the pit where the hair enters the skin and pass by layers and layers of boxy chambers stacked like nearly transparent bricks. The first layers seem dried out and dead, but deeper down the boxes are filled with fluid. Vague shapes

float inside. You realize you're looking at cells that make up the tissue of the prairie dog's skin.

You come to a giant tube stretching and branching among the skin cells. The walls of the tube are also made of cells, but these are thin and flat like flexible tiles. Looking through the cells that make up the tube—a blood vessel—you see membrane-bound sacs whizzing past like cars on a freeway. Red blood cells look a bit like life preservers with deep dents in the center. White blood cells look spiky.

On a whim, you hop aboard one of these red "life preservers." You've chosen a particularly bright red one, and off you go, bringing oxygen to another part of the prairie dog's body. The tubes become narrower and narrower, until you've reached the lining of the prairie dog's small intestine. Here's where the blood cell delivers its oxygen. You get off, too.

Cells at Work

These small intestine cells are busy. You watch as they produce enzymes and acids to break down food that the prairie dog has eaten. The chemical-producing cells are located right next to other types of cells, which have a different task: absorbing nutrients from the food. These nutrients pass through the membranes of the digestive cells until they reach the same narrow blood vessel where you first came in. Here the prairie dog's blood picks up the food and sends it to other parts of the body.

Maybe it's time you got out of this prairie dog. You close your eyes, and when you open them again, you see blue sky above. Tall stalks of . . . something . . . frame your vision—are they giant prairie dog hairs? No. You sit up and realize that you had fallen asleep among the tall grasses of the prairie.

Activity

BUILD YOUR OWN **Here's how you can study cells in both two and three dimensions.**

1. Slice into an onion and peel off part of an inner layer of skin. Study this layer under a hand lens or magnifying glass and describe what you see. Make a sketch.

2. Prepare one cup of plain gelatin (follow instructions on box). Pour the solution into a clear plastic baggie and refrigerate. When the gelatin has set, remove it from the refrigerator. Describe what you see and list the characteristics that make this "cell" different in structure from the onion cells.

It's a Small, Small World

One-celled organisms may be tiny, but they are everywhere. There is hardly a place on Earth that isn't inhabited by one kind of microorganism or another. We have identified some species, but there are many others we haven't yet discovered. Each species is specially adapted to live in its environment. And some of these environments aren't exactly easy places to live, either. Here's a survey of some one-celled organisms and the places they call home.

❸ Mid-Atlantic Coast: A Cell to Die for
In recent years, a free-floating alga called *Pfiesteria piscicida* has killed more than 1 billion fish all over the world. No wonder its species name in Latin means "fish killer"! Since the early 1990s, *Pfiesteria piscicida* has threatened marine life along the coasts of Maryland and North Carolina. There it blooms in large colonies known as "red tides."

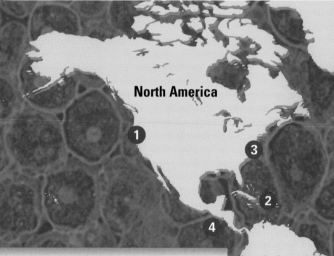

North America

South America

❶ Northern California: Superbug!
Scientists have discovered primitive single-celled bacteria at Iron Mountain in Northern California. The bacteria live in a hot, damp underground environment and thrive at the highest level of acidity. The surrounding rocks supply sulfur, iron, and manganese—all energy sources for the bacteria.

❷ Dominican Republic: Ancient Signs of Life
In the Dominican Republic, two scientists identified spores (a dormant form of bacteria) from a bacterium in the stomach of a bee almost 40 million years old. The bee was trapped in a piece of amber, a fossilized form of pine-tree resin. The scientists restored the ancient spores to life by placing them in a petri dish filled with nutrients.

❹ Costa Rica: A Beautiful Relationship
Sloths and algae have a good working arrangement. The algae live in the fur of the sloth (left), a mammal in the rain forests of Central America. Thanks to their blue-green color, the algae give the sloth a greenish tinge, which camouflages it from such predators as harpy eagles. Because sloths live high in the trees, algae in their fur get better exposure to sunlight than what's available on the forest floor.

Algae give this sloth its greenish color.

⑧ Africa: A Pest in a Pest

People assume mosquitoes are to blame for the disease known as malaria, but the real culprit is a single-celled parasite called *Plasmodium.* Mosquitoes can spread malaria by biting an infected person or animal. They suck up the malaria parasite in the blood and spread it to the next person or animal they bite. *Plasmodium* thrives in swampy tropical areas all over the world.

⑤ Northern Siberia: The Extremophiles

Russian scientists are studying cryophiles, microorganisms that live in the permafrost of northern Siberia and other intensely cold environments. Their ability to live in such extreme conditions gives the microorganisms their name.

⑥ East Coast of China: Red Tide Strikes Again!

When certain species of algae containing reddish pigment bloom near the ocean's surface, the water looks red. This can be a sign of the dreaded red tide, algae deadly to marine life. These organisms start out as cysts on the ocean floor. When the Sun warms the water, the cyst will break open and change into a one-celled organism that can swim. This cell rapidly divides, until it has produced several hundred algae within a couple of weeks. As the numbers of these cells increase, so do the dangers of the red tide.

Europe

Asia

Africa

Australia

⑦ Australia: Woodwinds

Deep inside a termite's stomach and intestines lives a wood-digesting cell. This organism produces enzymes to help the termite digest its diet of wood. In Australia, Aborigine musicians bury pieces of wood so the termites will chew them up. From the hollowed-out wood, the musicians make a didgeridoo (dij-er-ee-DOO), a traditional wind instrument (right).

Australian musicians rely on wood-digesting cells to make this traditional woodwind.

Activity

EXTRATERRESTRIAL LIFE? The more we know about "extremophiles" and other microbes that survive in extremely cold or hot environments, the more we might understand how life could exist on other planets. What if you were assigned to study environments elsewhere in our solar system—Mars, Venus, and Europa (one of Jupiter's moons)? Do some research and then choose which single-celled organisms you would bring on each mission. Present your research proposal to the class. Explain which cells you'd bring, and why.

The Case of the Basil Bandit

"I just can't believe it,"

said Mrs. Celia Lowes, shaking her head. "One of my prize plants is missing from the greenhouse. Did any of you see or hear anything unusual in the neighborhood this morning?" Mrs. Lowes—or Grandma Lowes, as most people called her— always happened to have freshly baked cookies in her house, which was a favorite stop for students on the way home from Louis Pasteur Middle School.

Gene Adams, Carrie Oats, and Amy Boyd had just stopped by after Saturday afternoon soccer practice. They all looked shocked by the news.

"Wow," said Carrie through a mouthful of peanut butter cookie. "I didn't notice anything. I was helping my mother with the laundry this morning before soccer."

"I didn't notice anything either," said Gene. "I worked with my uncle this morning in his new restaurant, the Hot Potato, same as I did two other nights this week."

"What's your job there?" asked Carrie.

"Basically, peeling potatoes and making French fries. It's good pocket money, but working at night is really cutting into my homework time."

"Well, I slept really late today," Amy said. "What kind of plant was it, Grandma Lowes?"

"One of my sweet basils."

"Well," Carrie mused, "at our school science fair we're participating in a plant project. We're all growing various plants and vegetables from seeds in different environments to see which ones will grow the biggest and healthiest—"

"We're learning all kinds of things about plants," Amy interrupted. "Did you know, Grandma Lowes, that a cucumber is actually a fruit, not a vegetable?"

"Well, I never!" said Grandma Lowes. "So what does that make a potato?"

"A potato's a vegetable," Gene said. "It stores extra food in the plant's underground stem. When it comes to potatoes, I'm the expert."

"Anyway, as I was saying," Carrie said, turning to Grandma Lowes, "I bet someone wants to use your basil for their science fair project."

"Let's check out the scene of the crime," said Amy.

Grandma Lowes led the children through her house to the backyard, to a glass greenhouse. "It must have happened this morning while I was busy in the kitchen baking bread. When I came out here this afternoon to get some fresh dill, that's when I noticed the missing sweet basil."

Amy, Gene, and Carrie looked around the greenhouse. Gene got sidetracked by the fragrant rose bushes and ended up pricking his finger badly. He followed Grandma back to the house for a bandage.

Meanwhile, Carrie looked for footprints but didn't find much because the ground was covered with gravel. Then Amy noticed something. "Do you normally keep the windows open in here?" she asked Grandma Lowes, who had returned. Gene was behind her, his finger neatly bandaged.

"I didn't open that window," Grandma Lowes replied. "How strange."

Carrie's face lit up. "That's a clue! Do you have some plastic wrap that we can use?" she asked.

"Well, of course. I'll go get some."

Grandma Lowes returned shortly with a sheet of plastic wrap, and Carrie divided it among the three of them. "Okay, Amy and Gene, let's see if we can collect smears of evidence on any surfaces the perpetrator might have touched. Maybe we'll get some cellular proof and look at it under the microscope in Mr. Hooke's science lab. He's at school today setting up for the science fair."

"Great idea!" said Gene. "I'll check the doorknob—it's made out of glass."

"Yeah," agreed Carrie. "It looks like there's a smudge on this open window. Maybe that's evidence."

Each of them pressed the cellophane wrap against the glass surfaces. Fifteen minutes later, the three sleuths caught up with Mr. Hooke in the school science lab. "We need your help to solve a crime!" shouted Carrie. She explained what had happened.

"Well, kids, let's see what we can find if we magnify things a bit," said Mr. Hooke, as he set each of them up with a microscope.

"This sample from the inside doorknob has a lot of red in it," Amy remarked. Mr. Hooke took a look it. "Hmm. That looks like blood—and fresh, too."

Then he went over to Gene. "What did you find?" he asked.

"I took this sample from the outside doorknob. These cells look really weird—like little bubble globules all stuck together."

"That looks like it could be baker's yeast," suggested Mr. Hooke.

"Look at this, Mr. Hooke," Carrie called. "My sample is green and also has a lot of white in it."

"That green is chlorophyll. You've certainly got a plant sample there," said Mr. Hooke. "And these white samples are very starchy cell walls. Plants make starch during photosynthesis. Some plants also store it in their stem and roots."

"That's obviously the basil that was stolen," said Carrie. "Looks like we're not really getting anywhere."

But Mr. Hooke was still looking at Carrie's sample. "Wait a minute. It looks like you've got two separate plant samples here. One of these has considerably more starch."

"Hmmm," said Carrie, thinking. "Starch . . . you know, I think I know who did it."

What did Carrie guess? Who do you think took Grandma Lowes's plant, and why? Answer the questions in the clue box below to help solve the mystery.

Clues

Use these clues ...

1. What kinds of foods come from plants heavy in starch?

2. What is baker's yeast used for?

3. Why were the cell samples from the outside doorknob, the inside doorknob, and the window different?

4. Where did the blood come from?

Answers on page 32

CELL-u-LITE

Bacteria Factoids

● There are more than 1 billion (1×10^9) bacteria in one teaspoon of topsoil.

● Bacteria living inside you outnumber your own cells. (You have about 50 million million cells in your body, so that's a lot of bacteria!)

● Some bacteria are so tough they can survive radiation a thousand times more intense than fatal radiation levels for humans.

● Bacteria can survive during long periods of starvation. Most shrink to less than a thousandth of their normal volume, as they slowly consume their internal food supply. These hungry microbes are called dwarf bacteria, or "ultramicro-bacteria."

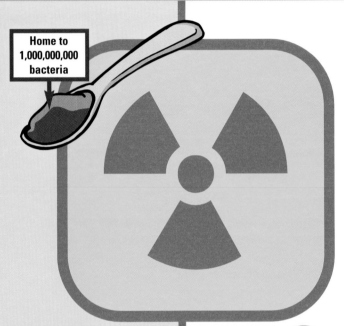

Home to 1,000,000,000 bacteria

Party Time

Q: What's another name for a bacteria party?
A: A cell-abration.

Q: How were the guests invited?
A: They were contacted on their cell phones.

Q: Where did all the bacteria go when neighbors complained that the party was getting too loud?
A: Down in the cell-ar.

Q: What did they do with all the leftover food?
A: They wrapped it up in cell-ophane

THE MEANING OF LIFE

Cell terms have interesting origins. Here's a sample:

Membrane: Latin for "thin skin"

Nucleus: Latin for "nut" (think of a cherry pit)

Cilia: Latin for "eyelashes"

Protozoa: Greek for "first animal"

One-celled organisms

MICROBES TO THE RESCUE

We're used to thinking about microbes as the cause of diseases and infections. But did you know that many one-celled organisms are actually beneficial to other forms of life? Check out these microbial nuggets:

- Yeast are fungi that produce carbon dioxide. They make bread dough rise.

- Some bacteria in your intestines produce Vitamin B, which you need to be healthy and strong.

- Half of the oxygen in our atmosphere comes from algae and bacteria.

- Some forms of bacteria consume harmful chemicals from oil spills.

Yeast

Algae

The Case of the Missing Microscopes

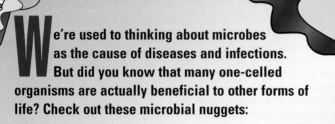

Microscopes made by Anton van Leeuwenhoek are very valuable today. During his lifetime he built more than 400 of them, including some out of silver. After he died in 1723, the silver scopes were given to the Royal Society of London, and the remaining scopes were sold at auctions. Yet only nine of these can be accounted for today. What happened to the hundreds of others? No one knows. If you see a Leeuwenhoek microscope at a museum, it's probably a copy, because the original nine are so valuable that they are usually kept in safes. But every now and then, a microscope said to be an authentic Leeuwenhoek comes to light. Ultimately, it turns out to be a fake. But if all the real ones are kept hidden away in safes, how could so many fakes keep turning up? Could it be that there are more surviving Leeuwenhoek microscopes around than we thought? To make a convincing fake, you need to see the original, right? Maybe there's a genuine Leeuwenhoek microscope hidden in someone's attic in your very own hometown . . . you never know.

Some Tough Cells

The world is full of an amazing diversity of organisms, and the smallest unit of every one of these is a cell. To appreciate the incredible variety of cells out there, get your classmates together and study this painting. It shows a natural scene with different kinds of plant and animal life.

Divide into small groups. Each group works with one item from the following list of subjects shown in the painting:

- Human beings
- Bird
- Tree
- Bush
- Flower
- Mammal

Exotic Landscape with Lion and Hunters, Henri Rousseau (1844–1910)

Over the next two days, see what you can find out about your subject and all the different kinds of cells it has. When the class gets together again, each group can present a list of all the different varieties of cells they found. Groups can write their lists on the blackboard, or make posters for the class to see.

Next, spend one or two days finding out about the functions of each of the cells you listed. These might be one-celled organisms that are specially adapted to their environments, or specialized cells within a larger multi-celled organism. Share your results with the other groups and have a class discussion: Did some groups come across the same kinds of cells? Are there some cells unique to just one group? If so, why?

Ready for the ultimate challenge? Enter this or any other science project in the Discovery Young Scientist Challenge. Visit *discoveryschool.com/dysc* to find out how.

ANSWERS

Scrapbook, page 17

1) 7×10^{-5} m

2) 5×10^{-6} m to 8×10^{-6} m

3) 5×10^{-1} m

4) 5×10^{-7} m to 1.5×10^{-6} m

Solve-It-Yourself Mystery, pages 28–29

Gene Adams, the chef's helper, is the burglar. He climbed into Mrs. Lowes's greenhouse window, after working at his uncle's restaurant peeling potatoes, and left starchy deposits on the window. Gene later admitted that he had been so busy working to make money that he hadn't done his science fair project. He hoped to win the science fair with Mrs. Lowes's sweet basil, and he planned to return it afterwards. The red sample on the outside doorknob was blood from Gene's pricked finger. The baker's yeast on the inside doorknob was from Grandma Lowes's hands—she said she had been baking bread. Once Gene got caught, however, he returned the plant to Grandma Lowes, who promised to give him some growing tips, so that he could have success growing his own plants.